VACCINE FOR KIDS

Facts you need to know about the vaccine

Jack Laurie

Jack Laurie

CopyRight©2021 Jack Laurie

All Right Reserved

TABLE OF CONTENT

INTRODUCTION

CHAPTER ONE

WHAT IS THE RECOMMENDED COVID-19 VACCINE FOR CHILDREN?

CHAPTER ONE

TEN FACTS ABOUT VACCINES AND THEIR ROLE IN GLOBAL HEALTH

INTRODUCTION

Coronavirus disease 2019 (COVID-19) is a contagious disease caused by coronavirus 2 that causes severe acute respiratory syndrome (SARS-CoV-2). The first case was discovered in Wuhan, China, in December of this year. Since then, the disease has spread over the world, resulting in a pandemic.

Fever, cough, headache, exhaustion, breathing difficulty, and loss of smell and taste are common symptoms of COVID-19. Symptoms can appear anywhere from one to fourteen days following viral contact. At least a third of those infected do not show any signs or symptoms. Most patients (81%) develop mild to moderate symptoms (up to mild pneumonia), whereas 14 percent develop severe symptoms (dyspnea, hypoxia, or more than 50% lung involvement on imaging), and 5% develop critical symptoms (respiratory failure, shock, or multiorgan dysfunction). Severe symptoms are more likely to emerge in older adults. Some persons continue to have a variety of symptoms (long COVID) months after recovery, and organ damage has been seen. Long-term research is being conducted to better understand the disease's long-term impact.

COVID-19 is spread by inhaling contaminated air that contains droplets and small airborne particles harboring the virus. Breathing them in is most dangerous when individuals are close together, although they can be inhaled over greater distances, especially indoors. Transmission can also happen if contaminated fluids are splashed or sprayed in the eyes, nose, or mouth, as well as via contaminated surfaces. People can be contagious for up to 20 days and can spread the infection even if they are symptomless.

To diagnose the condition, a variety of diagnostic approaches have been devised. The nucleic acid of the virus is detected by real-time reverse transcription polymerase chain reaction (rRT-PCR), transcription-mediated amplification (TMA), or reverse transcription loop-mediated isothermal amplification (RT-LAMP) from a nasopharyngeal swab as the standard diagnostic approach.

Several COVID-19 vaccines have been licensed and are being delivered in nations that have begun mass vaccination efforts. Other precautions include physical or social separation, quarantining, indoor ventilation, cough and sneeze covers, hand washing, and keeping unclean hands away from the face. In public places, the use of face masks or coverings has been suggested to reduce the risk of transmission. While efforts are underway to create antiviral medicines, the primary treatment is symptomatic. Treatment for symptoms, supportive care, isolation, and experimental approaches are all part of the management process.

COVID-19 is caused by infection with the SARS-CoV-2 virus strain.

COVID-19 is a virus that is transmitted from person to person.

COVID-19 is disseminated by the respiratory system, which includes bigger droplets and aerosols.

Persons inhale droplets and small airborne particles (that form an aerosol) that infected people exhale while they breathe, talk, cough, sneeze, or sing. When infected people are physically close, they are more likely to spread COVID-19. Infection, on the other hand, can spread across longer distances, especially inside.

Infectivity might occur 1-3 days before symptoms appear. Even if they are asymptomatic or pre-symptomatic, infected individuals can spread the disease. The peak viral load in upper respiratory tract samples usually occurs around the time of symptom onset and diminishes after the first week. According to current research, viral shedding and infectiousness can last up to 10 days for people with mild to moderate COVID-19 and up to 20 days for people with severe COVID-19, including immunocompromised people.

Infectious particles come in a variety of sizes, from aerosols that stay suspended in the air for a long time to bigger droplets that stay airborne or fall to the ground. Furthermore, COVID-19 research has reshaped our understanding of how respiratory viruses spread.

The largest droplets of respiratory fluid do not go far and can infect mucous surfaces of the eyes, nose, or mouth if breathed or landed on them. Aerosol concentrations are highest when humans are physically near together, making viral transmission simpler. However, airborne transmission can occur over longer distances, especially in poorly ventilated areas, where small particles can remain suspended in the air for minutes to hours.

Only 10 to 20% of persons are responsible for the spread of the disease, therefore the number of people infected by one infected person fluctuates. It usually spreads in clusters, with illnesses linked to an index case or a specific geographic place. Super spreading occurrences, in which a single individual infects a large number of people, are common in these situations.

Illustration of the SARSr-CoV virion in virology coronavirus 2 SARS-CoV-2 is a new severe acute respiratory syndrome coronavirus. It was first found in three persons with pneumonia who were linked to a Wuhan cluster of acute respiratory sickness cases. In nature, all of the structural properties of the novel SARS-CoV-2 virus particle can be found in similar coronaviruses.

The virus is eliminated outside the human body by household soap, which breaches the virus's protective bubble.

SARS-CoV-2 is a close relative of SARS-CoV. It is assumed to have a zoonotic (animal-borne) origin. The coronavirus genetically clusters with the genus Betacoronavirus, in the subgenus Sarbecovirus (lineage B), together with two bat-derived strains, according to genomic study. At the complete genome level, it is 96 percent identical to other bat coronavirus strains (BatCov RaTG13). Membrane glycoprotein (M), envelope protein €, nucleocapsid protein (N), and the spike protein are the structural proteins of SARS-CoV-2 (S). The M protein of SARS-CoV-2 is 98 percent similar to the M protein of bat SARS-CoV, 98 percent similar to pangolin SARS-CoV, and 90 percent similar to the M protein of SARS-CoV, but only 38 percent similar to the M protein of MERS-CoV. The M protein's structure mimics that of the sugar transporter SemiSWEET

Variants of the SARS-CoV-2 virus

SARS-CoV-2 variants are the main topic of this article.

SARS-CoV-2 variations are divided into clades and lineages. By GISAID, Nextstrain, and Pango, the WHO has built nomenclature systems for designating and tracking SARS-CoV-2 genetic lineages in partnership with partners, expert networks, national authorities, institutions, and researchers. At this time, a WHO-convened expert committee has advised that variants be labeled using Greek Alphabet letters, such as Alpha, Beta, Delta, and

Gamma, with the argument that they will be easier and more practical to discuss by non-scientific audiences. GISAID splits the variations into seven clades, while Nextstrain divides them into five (19A, 19B, 20A, 20B, and 20C) (L, O, V, S, G, GH, and GR). Many circulating lineages are classified as B.1 using the Pango tool.

Several noteworthy SARS-CoV-2 mutations emerged in 2020. In Denmark, Cluster 5 has emerged among minks and mink producers. The cluster was determined to be no longer circulating among humans in Denmark as of February 1, 2021, after stringent quarantines and a mink euthanasia campaign.

As of July 2021, four dominant SARS-CoV-2 variants are spreading among global populations: the Alpha Variant (formerly known as the UK Variant and officially referred to as B.1.1.7), which was first discovered in London and Kent, the Beta Variant (formerly known as the South Africa Variant and officially referred to as B.1.351), the Gamma Variant (formerly known as the Brazil Variant and officially referred to as P.1), and the Delta Variant (formerly known as the (formerly called the India Variant and officially referred to as B.1.617.2).

Using whole genome sequencing, epidemiology, and modeling, researchers believe the B.1.1.7 lineage's Alpha variation VUI-

202012/01 (the initial variant under investigation in December 2020) spreads more easily than other strains.

Other diseases' effects

On March 3, 2021, it was reported that social distancing and the widespread usage of surgical masks and similar protective gear as a preventative measure against COVID-19 resulted in a decrease in the spread rate of the common cold and flu. So much so that between 30 November 2020 and 21 February 2021, sales of cough drink, throat lozenges, and decongestants in the United Kingdom were roughly half of what they were a year earlier. To now, Public Health England has reported no incidents of flu in the year 2021, and sales of Vitamin D to help build immunity have increased by 89 percent.

Vaccines against the COVID-19 virus.

To halt the COVID-19 pandemic, equitable access to safe and effective vaccines is crucial, therefore seeing so many vaccines in research and testing is really encouraging. WHO collaborates with partners to develop, manufacture, and distribute vaccinations that are safe and effective?

Safe and efficient vaccines are a game-changing tool, but we must continue to wear masks, wash our hands, ensure proper ventilation indoors, and physically separate ourselves from crowds for the foreseeable future.

Being vaccinated does not imply we can disregard caution and put ourselves and others at danger, especially since research into how well vaccines protect against disease, infection, and transmission is still ongoing.

However, immunization, not vaccinations, will be the key to halting the epidemic. We must ensure that vaccinations are distributed fairly and equally to all countries and that each country receives and can use them to safeguard its citizens, beginning with the most vulnerable

.

CHAPTER ONE

What is the recommended COVID-19 vaccine for children?

The US Food and Drug Administration (FDA) have granted emergency use authorization to the Pfizer-BioNTech COVID-19 immunization for children aged 12 to 15. This vaccine, now known as Comirnaty, has also been approved by the FDA to prevent COVID-19 in people aged 16 and up.

The COVID-19 vaccine from Pfizer-BioNTech requires two injections spaced by 21 days. In the event that more time is needed, the moment measurements can be provided up to six weeks after the main dose.

The Pfizer-BioNTech COVID-19 vaccine appears to be 100 percent effective in preventing COVID-19 in children aged 12 to 15. In people aged 16 and up, the antibody is 91 percent effective at predicting severe illness caused by COVID-19. Early research also suggests that the antibody is 96 percent effective in predicting serious illness caused by COVID-19 caused by the delta variant, which is currently the most common COVID-19 variant in the United States.

How did the FDA assess the safety and efficacy of the Pfizer-BioNTech COVID-19 vaccine for children aged 12 to 15 years?

The FDA looked at the results of a study involving over 2,200 children aged 12 to 15 from across the United States. The Pfizer-BioNTech COVID-19 vaccine was given to about half of the group. The inactive (placebo) shot was given to the other children.

In the 1,005 youngsters who received the Pfizer-BioNTech vaccination a week following the second dosage, study revealed no cases of COVID-19. There were 16 cases of COVID-19 among 978 children given the placebo. COVID-19 had never been identified in any of the children before. The findings indicate that in this age group, the vaccine is 100 percent effective in preventing COVID-19.

What are the risks associated with the Pfizer-BioNTech COVID-19 vaccine for children aged 12 to 15?

The Pfizer-BioNTech COVID-19 vaccination caused side effects in children aged 12 to 15 that were similar to those seen in adults aged 16 and up. The following are the most frequently reported adverse effects:

Pain at the area where the injection was given

Fatigue \headache

Chills

Muscle aches

Fever

Pain in the joints

Children, like adults, experience side effects that last one to three days. With the exception of injection site pain, more teenagers experienced these side effects after the second dosage of the vaccine. Many others, on the other hand, experience no negative side effects.

Following the administration of the COVID-19 vaccination, your child will be monitored for 15 to 30 minutes to evaluate if he or she develops an allergic reaction that necessitates treatment.

To avoid adverse effects, don't give your child an over-the-counter pain medicine before immunization. It's fine to give your child this medication once he or she has received the COVID-19 vaccine.

Can the COVID-19 vaccine have an effect on the heart?

Following mRNA COVID-19 vaccination, there has been a surge in recorded incidences of myocarditis and pericarditis in the United States, mainly among male teenagers and young adults aged 16 and older. Myocarditis is an inflammation of the heart muscle, whereas pericarditis is an inflammation of the heart's outer lining. These types of reports are uncommon. The Centers for Disease Control and Prevention (CDC) is looking into whether there is a link between the COVID-19 immunization and the outbreak.

The problem occurred more frequently after the second dosage of the COVID-19 vaccine, and typically within a few days following COVID-19 vaccination, according to the instances recorded. After obtaining medicine and resting, the majority of those who received prompt care felt much better. The following are some warning signs to look out for: Chest ache, Breathlessness, Feelings of a pounding, fluttering, or racing heart Seek medical help if you or your kid develops any of these symptoms within a week of receiving the COVID-19 vaccine.

Is there any research on the long-term consequences of the COVID-19 vaccination from Pfizer and BioNTech?

Because clinical studies for the COVID-19 vaccine only began in the summer of 2020, it's still unclear whether the immunizations will have long-term impacts. Vaccines, on the other hand, rarely have long-term impacts.

After receiving the second dose of the COVID-19 vaccine, a part of the children aged 12 to 15 in the ongoing Pfizer-BioNTech

COVID-19 vaccine research were observed for at least two months for safety.

Pfizer Inc. developed a safety monitoring plan as part of its initial request for emergency use permission of its COVID-19 vaccine in 2020. Adolescents who have received the COVID-19 vaccine will now be monitored as part of the strategy.

In addition, in the United States, all immunization providers are required to report major adverse events, such as allergic reactions, to the Vaccine Adverse Event Reporting System, a national program.

Why do children require a COVID-19 vaccine if serious sickness from COVID-19 is rare?

A COVID-19 vaccine can protect your child from contracting and spreading the COVID-19 virus. If your child contracts COVID-19, the COVID-19 vaccine may help him or her avoid becoming seriously ill.

A COVID-19 vaccine will also enable your child to resume activities that he or she may have been unable to do due to the pandemic.

How does the COVID-19 vaccine from Pfizer and BioNTech work?

Message RNA is used in the Pfizer-BioNTech COVID-19 vaccine (mRNA). For decades, scientists have been studying mRNA vaccines.

Coronaviruses have an S protein, which is a spike-like structure on their surface. COVID-19 mRNA vaccines instruct immune cells on how to make a harmless S protein fragment. Following vaccination, cells begin to produce protein fragments and display them on cell surfaces. When the immune system recognizes a protein, it starts developing an immune response and producing antibodies. Once the protein pieces have been created, the cells deconstruct the instructions and discard them.

The vaccine's mRNA does not reach the nucleus of the cell, where DNA is stored.

Is there a difference in the Pfizer-BioNTech vaccine's ingredients or dosing for people aged 16 and up versus children aged 12 to 15?

No. The vaccine's ingredients and dosing are the same for all age groups.

Is the Pfizer-BioNTech COVID-19 vaccine appropriate for all children?

Children under the age of 12 are not eligible for this vaccine at this time. Clinical trials with younger children are currently underway.

A child who has had a severe allergic reaction to any of the vaccine's ingredients should not receive it. If this is the case, your child may be eligible for a future COVID-19 vaccine.

Is it possible for a COVID-19 vaccine to give a child COVID-19?

No. COVID-19 vaccines being developed in the United States do not contain the live virus that causes COVID-19.

Is it possible for the COVID-19 vaccine to affect fertility or menstruation?

There is no evidence that any COVID-19 vaccines are linked to infertility.

After taking the COVID-19 vaccine, a small number of women have reported suffering transitory menstrual abnormalities. A small study also found that after receiving COVID-19, some women reported transitory menstrual abnormalities. It's unclear if exposure to COVID-19 or the COVID-19 vaccination causes these alterations. Additional investigation is required.

Keep in mind that a variety of factors might influence menstrual periods, including infections, stress, sleep issues and dietary or exercise changes.

How can children aged 12 to 15 get vaccinated against COVID-19?

For information on where your child can acquire the COVID-19 vaccine, contact your local health department, drugstore, or your child's doctor. Consider the following questions before scheduling a COVID-19 vaccination appointment for your child:

Is a parent or guardian required to attend the appointment?

What information will be required during the meeting?

Is there a limit on the number of family members who can attend the appointment, such as siblings?

When can a child get a COVID-19 vaccine, either before or after another immunization?

COVID-19 and other vaccines can be administered on the same day.

Due to the newness of COVID-19 immunizations, the CDC previously advised against receiving any other vaccines for two weeks prior to and after receiving a COVID-19 vaccine. Based on recent safety findings, the CDC has revised its advice.

What may my child do once he or she has received the COVID-19 vaccine?

After getting fully vaccinated, your child can resume activities that he or she may have been unable to undertake due to the epidemic.

Two weeks following the second dosage of the Pfizer-BioNTech COVID-19 vaccine, your child is deemed fully immunized.

If your child doesn't exhibit symptoms after a known exposure in the United States, he or she won't need to be quarantined or have a COVID-19 test, with few exceptions for specific conditions.

Remember that COVID-19 immunization protects the majority of people from becoming unwell with COVID-19. Speak with your kid's doctor if you have any questions or concerns about your child receiving the COVID-19 vaccine. He or she may be able to assist you in balancing the risks and rewards.

vaccine for kids

CHAPTER ONE

TEN FACTS ABOUT VACCINES AND THEIR ROLE IN GLOBAL HEALTH

1. Vaccines protect up to 3 million youngsters from dangerous infections each year. At the same time, one in every five children 19 million children worldwide — lacks access to vital immunizations.

2. The measles vaccine has saved over 21 million lives worldwide since its introduction in 2000. Measles was eradicated in the United States in the same year, thanks to widespread vaccination campaigns.

3. Polio cases have decreased from 350,000 per year to only 33 cases last year, thanks to immunization efforts.

4. Rotavirus is the most prevalent cause of severe, often fatal diarrhea in children, and it can be easily avoided with a vaccine.

5. A clinical trial for a new universal influenza vaccine is now underway, which would eliminate the need to update the flu vaccination every year and boost protection against evolving flu strains.

6. All licensed vaccinations have undergone extensive testing in clinical trials and are continually evaluated for quality after they are released to the market. Vaccine responses are usually mild and only last a few days.

Vaccines are regarded as a "great deal" in the field of global health. According to UNICEF, every dollar spent on vaccines returns $44 in economic and social benefits.

8. Vaccines will be critical in reaching the Sustainable Development Goals (SDGs), a global agenda established by 193 countries in 2015 to make the world a better place where no one is left behind. This includes reducing poverty (SDG 1), addressing global hunger (SDG 2), and enhancing good health and well-being (SDG 3). (SDG 3).

9. The World Health Organization is currently evaluating a novel malaria vaccine, which is one of the world's oldest and deadliest diseases, killing over 435,000 people each year, the majority of them are children living in malaria-prone nations in Sub-Saharan Africa.

Jack Laurie

Antimicrobial resistance (AMR), a serious and growing danger to global public health, will require vaccines.